Superhero Cora

Measuring with Inches

Kathleen L. Stone

Dedication

To my newest granddaughter Cora, who joins her brother and sisters in our family and in our hearts!

Enjoy all these books by Kathleen L. Stone

Penguin Place Value
A Math Adventure

Number Line Fun
Solving Number Mysteries

Riley the Robot
An Input/Output Machine

Mason the Magician
Hundreds Chart Addition

Katelyn's Fair Share Picnic
More Math Fun

Money Tree Mysteries
Adventures with Quarters

Alien Even and Alien Odd
A Math Space Adventure

Kenley's Line Plot Graph
Another Math Adventure

Matthew's Sunshine Bakery
Multiplication Arrays

Firefighter Gary's Fire Safety Rules

Samantha's Search
3D Shapes

Grandma's Quilts
Fun with Fractions

Daniel's Day of Multiplication
Multiplication with Equal Groups

More Penguin Place Value
Hundreds, Tens, and One
s

Tick Tock Telling Time
Time to the Hour and Half Hour

Gavin the Gator
Greater Than and Less Than

Racecar Reecie
Elapsed Time

Superhero Cora
Measuring with Inches

From My Quilted Heart to Yours
Heart Warming Quilts and Heart Healthy Recipes for Your Loved Ones

From My Quilted Heart to Yours Book 2
Quilts and Blocks from the Children's Book, Grandma's Quilts

Children, Fire, and Intervention
Creating a Program that Saves Lives and Communities

Math Superhero

I'd like you to meet Cora
She's a math superhero.
She knows that when you measure
You should always start at *zero*.

No matter where she goes
She always has her ruler.
She loves finding things to measure.
To her there's nothing cooler.

Take a close look at her ruler.
Each long mark stands for *one* inch.
It's quite a handy tool
That makes measuring a cinch!

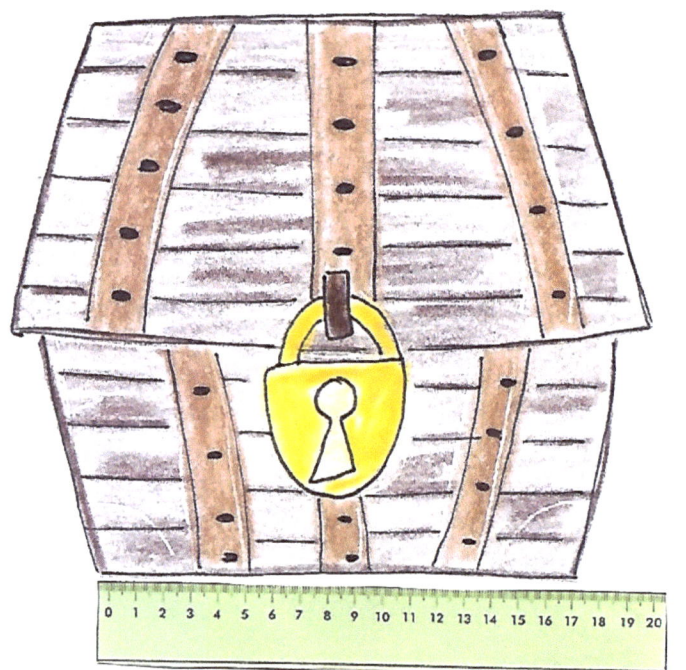

She always puts her ruler
Next to what she wants to measure.
Finding the correct measurement
Is just like finding a buried treasure.

Measure like this ...

<u>not</u> like this!

"One word of advice,"
Cora tells her little friend.
"Make sure your ruler and what
you're measuring
Both line up at one end."

Today Cora is measuring animals
Living in her backyard.
She'll show you how to do it.
It really isn't hard.

There's a little green worm
On the branch of that apple tree.
How many inches long is it?
Cora says, "It's a little over *three*."

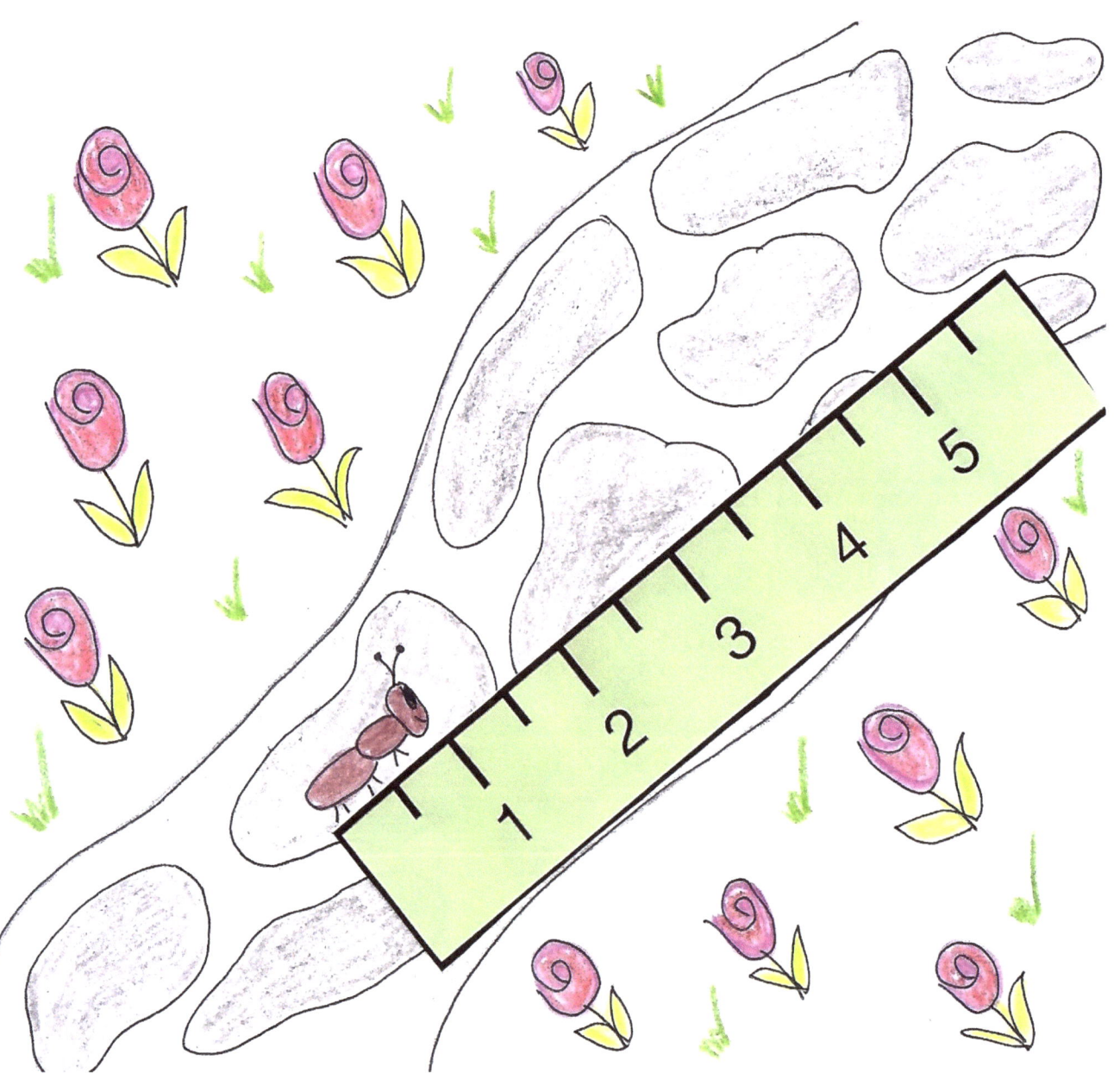

Next, Cora uses her ruler to measure
A tiny ant sitting in the sun.
How many inches long is she?
The ruler shows us *one*.

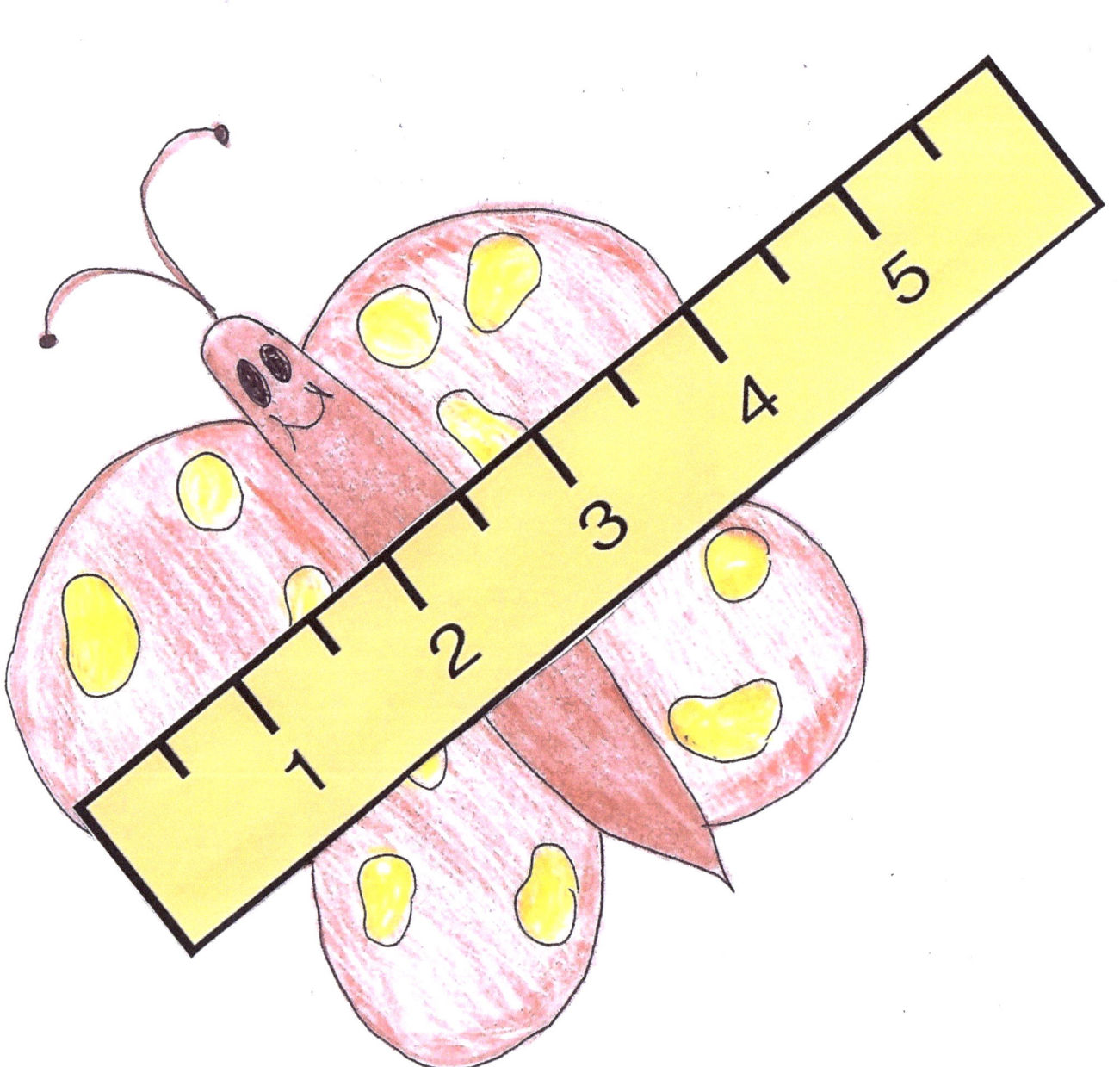

This beautiful spotted butterfly
Loves to flutter and to soar.
How many inches wide is it?
When we measure we see it's *four*.

How many inches long
Is that cat on Cora's gate?
Use your ruler carefully.
Did you find out that it's about *eight?*

Cora sees a garter snake
Hiding in a pile of leaves and sticks.
How many inches long is it?
The ruler shows that it's about *six*.

Now it is your turn
To find things around you to
measure.
You'll find that using this handy tool
Really is a pleasure!

Measuring with Inches

Measurements are a part of our daily lives whether we are measuring food, time, or objects. Children should be introduced early to the different types of measurements and tools that we use. Concrete, hands-on experiences using real rulers should always be encouraged to help develop this skill in children. I have found that my students learn even better when they are asked to estimate a length, width, or height of an object first and then given the opportunity to actually measure it to check their answer.

Enrichment Activities

Measure Hunt

Materials needed

rulers, tape measure, etc.
variety of "real life" objects

- ♥ Give children a measurement (for example, *three inches*) and have them find objects that they think are that long (tall or wide) and then have them measure to check their estimate.

- ♥ Provide lots of different measurements … *find something that is more than 12 inches long, less than 6 inches tall,* etc.

Measuring with Partners

Preparation of materials
- ♥ cut file folders in half
- ♥ glue picture* and question on the cover
- ♥ glue the correct answer inside the file folder

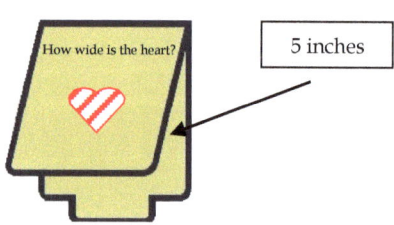

Playing the game
- ♥ place file folders around the room in easy to access areas
- ♥ Partner A uses ruler to measure the picture on the cover of the folder and states measurement. If Partner B agrees, they open the folder to check their answer
- ♥ partners move to another folder and this time Partner B measures the picture and Partner A will agree or disagree before checking inside the folder for the correct answer
- ♥ play continues until children have worked through all the folders

* old workbook pages and online clipart are great resources for pictures

ABOUT THE AUTHOR

Kathleen Stone is a National Board Certified educator and is currently teaching second grade. She loves spending time with her family. She and her husband Gary live in the Olympia area. When not teaching, Kathleen enjoys traveling with her husband, quilting, and reading by the lake. She also volunteers with West Thurston Regional Fire as a Youth Firesetter Interventionist, working with children and their families who have been involved in fire setting situations.

Math is all around us
No matter where you turn
Open your mind to the wonders of math
And all that you can learn

www.ingramcontent.com/pod-product-compliance
Lightning Source LLC
Chambersburg PA
CBHW041302180526
45172CB00003B/932